Eco-
Journey

EXPLORING
SEASHORES

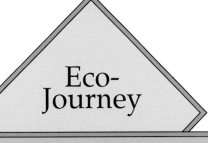

Eco-Journey

EXPLORING
SEASHORES

Barbara J. Behm
Veronica Bonar

Gareth Stevens Publishing
MILWAUKEE

For a free color catalog describing Gareth Stevens' list of high-quality books, call 1-800-341-3569 (USA) or 1-800-461-9120 (Canada).

ISBN 0-8368-1067-8

North American edition first published in 1994 by
Gareth Stevens Publishing
1555 North RiverCenter Drive, Suite 201
Milwaukee, WI 53212, USA

Photographic acknowledgments
The publishers wish to acknowledge, with thanks, the following photographic sources:
t = top *b* = bottom
Cover: Bruce Coleman Ltd.; Title page: Laurie Campbell/NHPA; pp. 6, 7 J. S. Gifford/NHPA; 8, 9*t* Bruce Coleman Ltd.; 9*b* Stephen Dalton/NHPA; 10, 11*t*, 11*b*, 12, 13*t*, 13*b*, 14 Bruce Coleman Ltd., 15*t* Stephen Dalton/NHPA; 15*b* Jeff Goodman/NHPA; 16 Anthony Bannister/NHPA; 17*t* Bruce Coleman Ltd.; 17*b* Laurie Campbell/NHPA; 18*t* Bruce Coleman Ltd.; 18*b* Melvin Grey/NHPA; 19, 20*t*, 20*b*, 21 Bruce Coleman Ltd.; 24*t* Jeff Goodman/NHPA; 24*b* Henry Ausloos/NHPA; 25 Bruce Coleman Ltd.; 26 Robert Harding Picture Library; 27 Bruce Coleman Ltd.

Printed in the United States of America

1 2 3 4 5 6 7 8 9 99 98 97 96 95 94

Title page:
Black-headed gulls swoop down to catch fish
in shallow inshore waters.

Contents

Words that appear in the glossary are printed in **boldface** type the first time they occur in the text.

Wind and water

Wind and water make cracks and ledges in the cliffs at the seashore.

▶ It is hard for plants and animals to live on seashore cliffs.

The top part of a beach is often sprayed with water. The middle area is covered and uncovered by each **tide**. The lowest area is always under the water.

▲ When the tide goes out, pools of water are left beside the rocks of the middle area of the beach.

The cliffs in spring

Pockets of soil build up in the cracks and hollows of cliffs. Plants grow there. In spring, this cliff face is covered by pink flowers.

▼ These flowers, called sea pinks, have long, deep roots that grow into the cracks in the cliff face.

The rocks high on the cliffs are covered with **lichens.** A lichen is made of two plants – a **fungus** and an **alga.** Huge colonies or groups of seabirds build their nests on these cliffs each spring.

▲ A young adder snake (right) warms itself in the springtime sun beside its mother. Adders live on the rocks near the top of the cliffs. They spend the winter in cracks in the rocks.

◀ Sea gulls called kittiwakes build nests on small ledges of cliffs.

9

Early summer

Some seabirds, such as oystercatchers, ringed plovers, and terns, nest on the shore. They dig shallow nests in the sand, well above the high-water mark.

▶ Tiny storm petrels nest in burrows they dig in the soil at the top of cliffs. Petrels have such short, weak legs that they can hardly walk on land. On water, they skim over the ocean waves. Their feet paddle on the surface as if they were walking on the water.

◄ These elephant seals are lying in the sun, safely above the high-water mark of the beach.

▼ The plant called horned-poppy has large blue-green leaves and golden flowers. Its "horns" are long, thin seed pods.

Plants such as sea kale, sea lavender, and sea campion grow in the sand at the beach. Rocks are crusted with black lichens. Seaweed grows in shallow, salty pools of water at the top of the beach.

The falling tide

When the tide goes out, the rocks of the middle beach are uncovered. **Marine** animals, such as periwinkles, barnacles, mussels, and limpets, can be seen as they cling to the rocks.

▶ Barnacles hang down from the rocks. They feed on the small animals and plants that wash their way.

◄ As the tide goes out, small fish and shrimplike animals called prawns are left behind in rock pools until the tide returns.

▼ Flat periwinkles have yellow, brown, orange, or striped shells. They live among the seaweed on the middle beach.

Brown seaweed called wrack grows on the rocks of the middle area of the beach. The wracks are anchored to the rocks by their roots.

Seashore predators

When the tide goes out, crabs and shrimps hide under rocks and seaweed from **predators**.

▼ The purple sandpiper will push its long beak into the sand, looking for crabs and worms.

◀ This snakelocks anemone has caught a prawn in its tentacles.

Crabs, snails, and sea anemones **prey** on smaller animals in the rock pools. Sea anemones have feelers called tentacles. When an animal touches an anemone, the tentacles wrap around the victim and inject a poison into it.

▼ A marine animal called a rock goby hides under a rock, waiting to catch its prey. Its coloring makes it hard to spot against pebbles in the rock pool.

In the rock pool

Pipefish hide among the weeds in rock pools when the tide is low. They feed on tiny plants and animals known as **plankton**.

▼ When a hermit crab has grown too big for the shell in which it lives, it must find a larger one. The crab quickly pulls its body out of the old shell and dives into its new home.

◄ When the tide falls, beadlet sea anemones pull in their tentacles. They look like blobs of jelly on the rocks. Fish called blennies sometimes leave the rock pools to perch on nearby rocks.

▼ Periwinkles, sea slugs, and sea urchins eat knotted wrack. Knotted wrack grows only in sheltered places. It cannot live where the waves are very strong.

Chameleon prawns hide in the weeds of rock pools. They can change color from green to red or brown to match the color of their background.

Hiding from others

At low tide, many tiny animals, like lugworms, sandmason worms, and peacock worms, hide beneath the sand. There, they feed on decaying plants and animals.

▲ A shore crab sits beside lugworm casts. Lugworms feed on plants and animals in the sand. The sand they have also eaten passes out of their bodies in the shape of their bodies. These are called casts.

▶ This black-backed gull has caught a crab.

◄ The air bubbles in bladder wrack make this seaweed float in the water at low and high tide.

During the day, shore crabs bury themselves in the sand or hide under stones or in rock pools. They come out at night when birds and other predators cannot see them.

Under the sand

Mollusks are animals that have soft bodies and a strong, muscular foot. Many mollusks hide in the sand, safe from predators.

▲ A sand wasp clears away sand from the entrance of its burrow. It has caught a caterpillar and will bury it in the burrow.

▶ Cockles are mollusks that live just below the surface of the sand. When the tide comes in, this cockle will push two tubes up through the sand to filter out particles of food from the water.

The lugworm lives in a U-shaped burrow under the sand. One end of the burrow fills with sand at each high tide. The worm then eats the sand. The grains of sand pass out of the worm's body, making a cast.

Sandmason worms also live beneath the sand. Each worm builds a tube-shaped home from grains of sand and bits of shells. When the tide is in, the worm climbs up the tube to eat food that drifts by.

21

At low tide

Starfish can be seen at low tide. Each starfish has five arms with a row of tube feet underneath each arm. Starfish are covered with bony plates.

▲ A sea urchin crawls over the rocks and eats seaweed called kelp.

▶ The starfish wraps itself around its prey. Its tube feet gradually pull open the shells of the victim.

◄ Sandhoppers can swim well in the sea, but they are easy prey for fish. When the tide comes in, the sandhoppers run ahead of the water.

Sandhoppers are **scavengers** that eat rotten seaweed or other dead plants or animals on the seashore. They feed on the beach at night to be safe from birds that eat them.

Just offshore

Many kinds of flatfish, such as sole and halibut, live at the bottom of the shallow waters just offshore. Flatfish have both eyes on the same side of their heads.

▲ A lobster's body is covered by a shell. It walks along the seabed on four pairs of legs. The fifth pair forms pincers that cut and crush its food. If a lobster loses a leg, it grows another one.

▶ Bottle-nosed dolphins are friendly animals that often swim near the coast.

◀ Small fish swim among the kelp. Many marine animals and their young feed on and find shelter among the kelp plants.

Kelp are seaweeds that grow in slightly deeper water. They weaken the force of waves on shore.

Winter storms

Autumn and winter storms crash against the rocks on the seashore. The wind and waves wear away the shoreline.

▼ A winter storm sends clouds of spray into the air as waves pound against the rocks. This violent action gradually changes the shape of the coastline.

◀ A large flock of golden plovers feeds near the shore.

Seabirds leave the cliffs to spend the winter in warmer waters or move inland to escape winter storms. Many birds that spend the breeding season in the Arctic will live at the seashore during winter.

More Books to Read

At the Beach. Eugene Booth (Raintree)

Beaches are for Kids! An Activity Book for Kids. Bobbi Salts (Double B)

How to Hide an Octopus: and Other Sea Creatures. Ruth Heller (Putnum)

Seashore Surprises. Rose Wyler (S and S Trade)

What's for Lunch? The Eating Habits of Seashore Creatures. Samuel Epstein (MacMillan)

Videotapes

Call or visit your local library to see if these videotapes are available for your viewing.

Octopus. Barr (Animal Family Series)

The Wonderful World of Disney: Pacifically Peeking. (an animated and live-action look at the geography, people, and wildlife in and near the Pacific Ocean)

Places to Write

Department of Fisheries and
   Oceans
200 Kent Street
Ottawa, Ontario K1A OE6

Sea World
Public Relations Department
1720 South Shores Road
San Diego, CA 92109

Sea World
Public Relations Department
7007 Seaworld Drive
Orlando, FL 32812

Interesting Facts

1. Flowers that grow on cliffs above the sea are not harmed by the salt in the spray from ocean waves.

2. The alga plant takes water from the fungus plant and uses it to make food for the fungus. These two plants need each other to live. Together, they form lichens.

3. The storm petrel lays just one large white egg in its burrow nest. It squirts a bad-smelling liquid at any being that goes too close to its nest.

4. When the tide comes in, peacock worms and sandmason worms unfold a fan of tentacles to catch floating pieces of food from the water.

5. The sea potato is actually an animal covered with soft, furry spines. It uses its spines to dig a burrow below the surface of the sand. It then pushes out long tentacles to breathe and catch food.

6. The shells of clams, mussels, and razorshells are in two halves, hinged down one side so the shells can open and close.

7. When a starfish eats, its stomach turns inside-out and then sucks up its food.

8. The starfish often swallows its prey whole. Later, it spits out any shells.

9. Sandhoppers have different kinds of legs for walking and swimming. Sandhoppers spend their time at the edge of the sea, half in and half out of the water.

10. The two largest kinds of sharks are the whale shark and the basking shark. They do not eat humans. They eat only plankton.

Glossary

alga: a tiny plant that is mostly aquatic, such as seaweed.

fungus: a plant that feeds on other plants or dead plants and animals.

lichens: sets of two plants that live together, a fungus and an alga, and help each other survive.

marine: anything that has to do with the sea.

mollusks: marine animals that have a soft body and a strong, muscular foot.

plankton: tiny plants and animals that live mainly at the surface of the sea.

predators: animals that kill other animals for food.

prey: an animal hunted by another animal for food.

scavengers: animals that feed on dead plants and animals.

tide: the rising and falling of the sea that usually occurs twice a day.

Index